Dimensions Mat
Tests 2B

MW01232994

Singapore Math Inc.

Published by Singapore Math Inc.

19535 SW 129th Avenue

Tualatin, OR 97062

www.singaporemath.com

Dimensions Math® Tests 2B

ISBN 978-1-947226-50-0

First published 2019

Reprinted 2020 (twice), 2021 (twice)

Printed in China

Acknowledgments

Editing by the Singapore Math Inc. team.

Design and illustration by Cameron Wray with Carli Bartlett.

Preface

Dimensions Math® Tests is a series of assessments to help teachers systematically evaluate student progress. The tests align with the content of Dimensions Math K–5 textbooks.

Dimensions Math Tests K uses pictorially engaging questions to test student ability to grasp key concepts through various methods including circling, matching, coloring, drawing, and writing numbers.

Dimensions Math Tests 1–5 have differentiated assessments. Tests consist of multiple-choice questions that assess comprehension of key concepts, and free response questions for students to demonstrate their problem-solving skills.

Test A focuses on key concepts and fundamental problem-solving skills.

Test B focuses on the application of analytical skills, thinking skills, and heuristics.

Contents

Chapter	Test	Page

BLANK

15 min **Score**

30

Test A

Chapter 8 Mental Calculation

Section A (2 points each)

Circle the correct option: **A**, **B**, **C**, or **D**.

1 $459 + 6 = 460 +$ ⬚

    A 4   **B** 7

    C 6   **D** 5

2 $360 + 80 =$ ⬚ $+ 40$

    A 400   **B** 300

    C 80   **D** 340

3 $100 - 82 = \boxed{}$

A 8

B 28

C 12

D 18

4 $405 - 98 = \boxed{} + 2$

A 405

B 407

C 305

D 307

5 What is 5 more than 179?

A 180

B 184

C 185

D 186

Section B (2 points each)

6 $99 + 37 =$ ☐

7 $705 - 7 =$ ☐

8 $320 - 70 =$ ☐

9 $621 + 30 =$ ☐

10 $996 - 97 =$ ☐

11 Complete the number pattern.

| 219 | 227 | | | 251 | |

12 Write >, <, or = in the ◯.

863 + 9 ◯ 869 + 3

13 There are 58 girls in a school hall.
There are 5 more boys than girls in the hall.
How many boys are there in the hall?

There are _____ boys in the hall.

14 200 people visited the museum on Sunday.
97 fewer people visited the museum on Monday than on Sunday.
How many people visited the museum on Monday?

_____ people visited the museum on Monday.

15 Chapa has $118.
She buys a coat for $99.
How much money does she have left?

She has $_____ left.

Name: _____

Date: _____

30

Test B

Chapter 8 Mental Calculation

Section A (2 points each)
Circle the correct option: **A**, **B**, **C**, or **D**.

1 $659 + 90 = 649 +$ ☐

   A 10   **B** 100

   C 190   **D** 80

2 $981 - 50 =$ ☐

   A 901   **B** 930

   C 951   **D** 931

3 $216 - 97 =$ ⬜

    A 19   **B** 119

    C 116   **D** 16

4 $98 + 300 = 400 -$ ⬜

    A 98   **B** 10

    C 2   **D** 8

5 What is 90 more than 289?

    A 299   **B** 389

    C 399   **D** 379

Section B (2 points each)

6 $99 + 899 = \boxed{}$

7 $776 + 97 = 876 - \boxed{}$

8 $47 = 100 - \boxed{}$

9 $108 - 10 = \boxed{}$

10 $880 + 80 = \boxed{}$

11 Write >, <, or = in the $\bigcirc$.

$34 + 5 + 720 \bigcirc 734 + 5 + 21$

12 Complete the number pattern.

| 99 | 198 | | | | 594 |

13 Carlos has $533.
Laila has $40 more than Carlos.
How much money does Laila have?

Laila has $_____.

14 A restaurant sold 100 tacos yesterday.
The restaurant sold 18 fewer burritos than tacos.
How many burritos did the restaurant sell?

The restaurant sold _____ burritos.

15 There are 928 boys at a school.
There are 7 more girls than boys at the school.
How many girls are at the school?

_____ girls are at the school.

Name: _____

Date: _____

25 min **Score**

30

Test A

Chapter 9 Multiplication and Division of 3 and 4

Section A (2 points each)
Circle the correct option: **A**, **B**, **C**, or **D**.

1 There are 7 tricycles.
How many wheels are there?

A 14   **B** 21

C 28   **D** 70

2 Emma put 18 cookies equally into 3 jars.
How many cookies are in each jar?

A 3   **B** 9

C 5   **D** 6

3 6 cars have _____ wheels.

A 24

B 12

C 18

D 20

4 40 ÷ 4 = [   ]

A 4

B 12

C 8

D 10

5 4 × 4 is 4 more than _____.

A 3 × 3

B 4 × 1

C 3 × 4

D 4 × 5

Section B (2 points each)

6 ☐ × 3 = 27

7 Circle products of 3.

| 25 | 18 | 30 | 20 | 8 | 9 |

8 Write × or ÷ in the ◯.

16 ◯ 4 = 4

9 There are 3 fish in each can.
How many fish are in 6 cans?

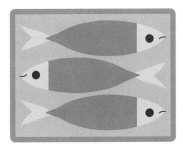

_____ fish are in 6 cans.

10 Aki has 12 cat treats.
She gives 3 cats the same number of treats.
How many treats does each cat get?

Each cat gets _____ treats.

11 1 bag of dog food weighs 6 kg.
How many kilograms do 4 bags of dog food weigh?

4 bags of dog food weigh _____ kg.

12 Kawai has a rope that is 12 meters long.
She cuts it into 4 pieces of equal length.
How long is each piece of rope?

Each piece of rope is _____ m long.

13 A teacher wants to give some students 4 counters each.
There are 9 students.
How many counters does the teacher need?

The teacher needs _____ counters.

14 One ferry ticket costs $4.
John pays $28 to buy a few ferry tickets.
How many tickets does he buy?

He buys _____ tickets.

15 A cook uses 3 eggs to make each omelet.
She has 30 eggs.

(a) How many omelets can she make?

She can make _____ omelets.

(b) How many eggs does she need to make 8 omelets?

She needs _____ eggs to make 8 omelets.

Name: _____

Date: _____

30

Test B

Chapter 9 Multiplication and Division of 3 and 4

Section A (2 points each)
Circle the correct option: **A**, **B**, **C**, or **D**.

1 There are 27 marshmallows.
Each stick has 3 marshmallows.
How many sticks of marshmallows are there?

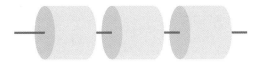

A 8   **B** 9

C 7   **D** 10

2 6×4 is _____ less than 7×4.

A 4   **B** 6

C 1   **D** 7

3 Cody's sister is 3 years old.

Cody is 3 times as old as his sister.

Cody is _____ years old.

A 6

B 12

C 9

D 10

4 $4 \times 10 = \boxed{}$

A $30 \div 3$

B 5×8

C 3×9

D 10×2

5 $36 \div 4 = \boxed{}$

A 3×4

B 3×9

C $27 \div 3$

D $16 \div 2$

Section B (2 points each)

6 $7 = 21 \div$ ☐

7 Write >, <, or = in the ◯.

$24 \div 3$ ◯ $40 \div 5$

8 $8 \times 4 = 20 +$ ☐ $=$ ☐

9 Circle the number that is both a product of 3 and a product of 4.

| 9 | 20 | 16 | 21 | 24 | 30 |

10 28 cm of tape is cut into pieces of the same length as the one shown here. How many pieces of tape are there?

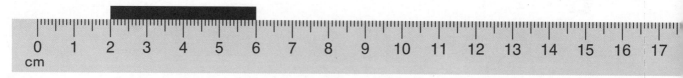

There are _____ pieces of tape.

11 Sydney wants to put 12 churros equally in 3 bags. How many churros will she put in each bag?

She will put _____ churros in each bag.

12 Muffins are sold in boxes of 4. Lily wants to bring 28 muffins to a party. How many boxes of muffins does she need to buy?

She needs to buy _____ boxes of muffins.

13 Carter has a ribbon that is 20 ft long.
He uses it to tie some presents.
Each present uses 3 ft of ribbon.

(a) How many presents can Carter tie with his ribbon?

Carter can tie _____ presents with his ribbon.

(b) How long is the leftover piece of ribbon?

The leftover piece of ribbon is _____ ft long.

14 There are 3 pieces of sushi on each plate.
Daren orders 7 plates of sushi.

(a) How many pieces of sushi does he get?

Daren gets _____ pieces of sushi.

(b) Each plate of sushi costs $4.
How much does Daren pay for the sushi?

Daren pays $_____ for the sushi.

15 A cook uses 3 eggs to make each omelet.

He has 16 eggs.

How many more eggs does he need to make 6 omelets?

He needs _____ more eggs to make 6 omelets.

Name: _____

Date: _____

30

Test A

Chapter 10 Money

Section A (2 points each)
Circle the correct option: **A**, **B**, **C**, or **D**.

1 What is the total amount of money?

A $6.25

B $6.36

C $6.31

D $6.30

2 Fifty dollars and seventeen cents is _____.

A $17.50

B $57.00

C $50.17

D $15.70

3 840¢ = ⬚

A $8.40

B $8.04

C $84

D $0.84

4 $9.59 − $3.20 = ⬚

A $9.27

B $6.57

C $6.30

D $6.39

5 $5.20 + $0.02 = ⬚

A $5.18

B $5.22

C $7.20

D $5.40

Section B (2 points each)

6 14¢ + ☐ ¢ = $1

7 _____ quarters make $3.25

8 Write >, <, or = in the ◯.

$5.05 ◯ 495¢

9 Arrange the amounts of money from least to greatest.

Use the pictures to answer questions 10–12.

burrito $6.75

sandwich $7.30

muffin $1.80

10 Henry buys a sandwich and a muffin.
How much does he spend?

Henry spends $_____.

11 Aaliyah pays for a muffin with a $5 bill.
How much change does she receive?

Aaliyah receives $_____.

12 Caleb has $5.25
How much more money does he need to buy a sandwich?

Caleb needs $_____ more to buy a sandwich.

13 Avery has 1 five-dollar bill, 1 quarter, 3 dimes, and 5 nickels.
How much money does she have?

Avery has $_____.

14 A cookie costs $0.99.
A muffin costs $3.55
How much do they cost altogether?

They cost $_____ altogether.

15 Logan has $3.12.
Dexter has 68¢ more than Logan.

(a) How much money does Dexter have?

Dexter has $_____.

(b) How much money do they have altogether?

They have $_____ altogether.

Test B

Chapter 10 Money

Section A (2 points each)

Circle the correct option: **A**, **B**, **C**, or **D**.

1 $1 – 18¢ = _____

A 1 dollar and 18 cents   **B** 82 cents

C 8 dollars and 20 cents   **D** 82 dollars

2 5 quarters and 6 pennies is _____.

A $1.25   **B** $1.85

C $1.06   **D** $1.31

3 Seven cents is the same as _____.

A $0.70

B $0.07

C $7.00

D $70

4 $4.23 − $0.39 = ⬚

A $4.62

B 33¢

C $3.84

D $8.13

5 45¢ + 75¢ = ⬚

A $1.20

B 30¢

C 84¢

D 125¢

Section B (2 points each)

6 6¢ + $3.94 = $ ⬚

7 Write >, <, or = in the ◯.

$4.20 + 92¢ ◯ $4.92 + 20¢

8 Circle the amount of money that is less than 3 one-dollar bills, 2 quarters, 2 dimes, and 3 pennies.

| $3.83 | $4.73 | $3.63 | $3.93 |

9 A bus ticket costs $2.75.
How much do 2 bus tickets cost?

2 bus tickets cost $_____.

Use the pictures to answer questions 10–12.

$9.05   dinosaur
$4.10   doll
$1.35   yo-yo
$4.15   monkey
$4.35   truck

10 Noah bought the doll, the yo-yo, and the truck.
How much did he spend?

Noah spent $_____.

11 Sasha pays $10.00 for one toy and receives 3 quarters and 2 dimes back.
Which toy did she buy?

Sasha bought the _____.

12 Amanda bought the toy that costs $4.90 less than the dinosaur.
Which toy did she buy?

Amanda bought the _____.

13 A notebook and a pen cost $8.00 altogether.
The notebook costs $6.35.

(a) How much does the pen cost?

The pen costs $_____.

(b) How much more does the notebook cost than the pen?

The notebook costs $_____ more than the pen.

14 Kim saves $3.95.
Ethan saves 3 quarters more than Kim.

(a) How much does Ethan save?

Ethan saves $_____.

(b) How much do Kim and Ethan save altogether?

Kim and Ethan save $_____ altogether.

15 Andy has $5.80.

He buys a pencil for $0.80 and a box of crayons for $3.99.

How much money does he have left?

Andy has $_____ left.

Name: _____

Date: _____

Test A

Chapter 11 Fractions

Section A (2 points each)

Circle the correct option: **A**, **B**, **C**, or **D**.

1 Which shape has 1 half of it shaded?

A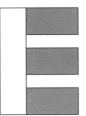

B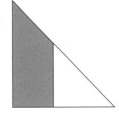

C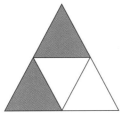

D

2 $\frac{5}{8}$ and _____ make 1 whole.

A $\frac{1}{8}$

B $\frac{3}{8}$

C $\frac{5}{8}$

D $\frac{3}{5}$

3 One-fifth means 1 out of _____ equal parts.

    A 1   **B** 6

    C 3   **D** 5

4 There are _____ one-tenths in $\frac{6}{10}$.

    A 1   **B** 4

    C 6   **D** 10

5 _____ is four-sixths.

    A $\frac{4}{6}$ **B** $\frac{1}{4}$

    C $\frac{1}{6}$ **D** $\frac{4}{10}$

Section B (2 points each)

6 Color $\frac{3}{10}$ of this figure.

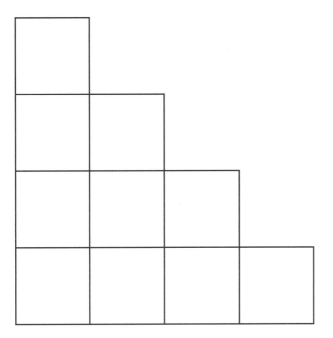

7 $\frac{3}{7}$ and $\boxed{\ \ \text{—}\ \ }$ make 1 whole.

8 What fraction of this figure is shaded?

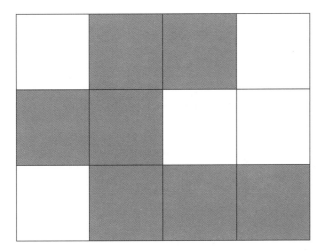

$\boxed{\dfrac{\ \ }{12}}$ of the figure is shaded.

9 Color $\frac{2}{9}$ of the bar.

10 Circle the largest fraction.

$\frac{1}{7}$ $\frac{1}{3}$ $\frac{1}{10}$

11 Circle the fraction that is smaller than $\frac{1}{6}$.

$\frac{1}{3}$ $\frac{1}{8}$ $\frac{1}{4}$

12 Arrange the fractions in order, beginning with the smallest.

$\frac{1}{4}$ $\frac{1}{2}$ $\frac{1}{7}$

_____, _____, _____

13 Carlos ate $\frac{1}{5}$ of a cake.
Ryan ate $\frac{1}{10}$ of the same cake.
Who ate more cake?

_____ ate more cake.

14 A pizza was cut into 8 equal slices.
Emma ate 2 slices, Sofia ate 1 slice and Mei ate 2 slices.
What fraction of the pizza did they eat altogether?

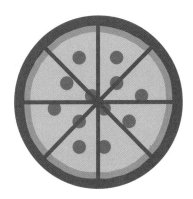

They ate _____ of the pizza altogether.

15 Chloe cut a lasagna into 9 equal pieces.
She ate 2 pieces.
What fraction of the lasagna was left?

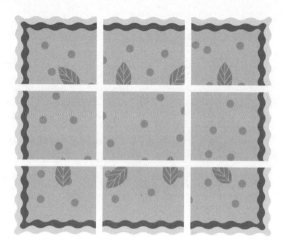

_____ of the lasagna was left.

25 min **Score**

30

Test B

Chapter 11 Fractions

Section A (2 points each)
Circle the correct option: **A**, **B**, **C**, or **D**.

1 Which shape has $\frac{1}{7}$ of it shaded?

A

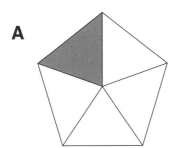

B

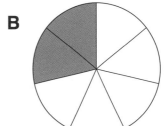

C

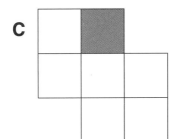

D

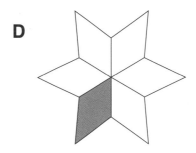

2 $\frac{1}{4}$ of a circle is larger than _____ of the same circle.

A $\frac{1}{2}$

B $\frac{1}{6}$

C $\frac{1}{3}$

D 1

3 There are _____ one-eighths in 1.

A 1

B 4

C 6

D 8

4 What fraction of the bar is shaded?

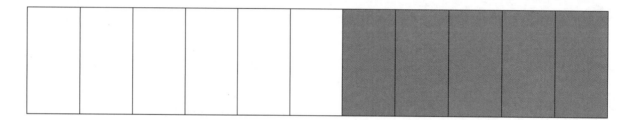

A $\frac{5}{11}$

B $\frac{4}{11}$

C $\frac{4}{10}$

D $\frac{8}{11}$

5 $\frac{3}{5}$ is _____.

A one-third

B three-sixths

C three-fifths

D one-fifth

Section B (2 points each)

6 Color $\frac{3}{10}$ of the figure.

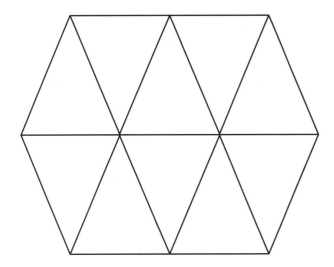

7

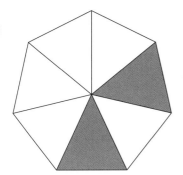

$\boxed{\dfrac{2}{}}$ and $\boxed{\dfrac{}{7}}$ make 1 whole.

8

There are 8 one-ninths in $\boxed{\dfrac{}{}}$.

9 Circle the smallest fraction.

$\frac{1}{4}$ $\frac{1}{2}$ $\frac{1}{6}$

10 Circle the fraction that makes 1 whole with five-tenths.

$\frac{1}{5}$ $\frac{1}{10}$ $\frac{5}{10}$

11 Arrange the fractions in order, beginning with the largest.

$\frac{1}{10}$ $\frac{1}{5}$ $\frac{1}{8}$ $\frac{1}{7}$

_____, _____, _____, _____

12 Circle the fraction that is larger than $\frac{1}{3}$.

$\frac{1}{4}$ $\frac{1}{9}$ $\frac{1}{2}$

13 Kyle ate $\frac{1}{5}$ of a melon in the morning.

He ate $\frac{1}{6}$ of the same melon in the afternoon.

Did he eat more of the melon in the morning or in the afternoon?

Kyle ate more of the melon in the _____.

14 John cut a pie into 10 equal pieces.

He gave 4 pieces to his neighbors.

What fraction of the pie did he give to his neighbors?

John gave _____ of the pie to his neighbors.

15 $\frac{1}{5}$ of the crayons in a box are red.

$\frac{1}{4}$ of the crayons in the box are blue.

Are there more red crayons or blue crayons in the box?

They are more _____ crayons in the box.

Continual Assessment 3

Section A (2 points each)

Circle the correct option: **A**, **B**, **C**, or **D**.

1 556 + 99 = ☐ + 100

   A 599   **B** 555

   C 557   **D** 455

2 237 − 8 = ☐

   A 245   **B** 227

   C 229   **D** 235

3 A cat weighs about 10 _____.

    A lb   **B** g

    C cm   **D** ft

4 There are 32 wheels.
Each car has 4 wheels.
How many cars are there?

    A 6   **B** 7

    C 8   **D** 32

5 The digit in the tens place of 481 is _____ more than the digit in the hundreds place.

    A 7   **B** 8

    C 3   **D** 4

6 7×3 is _____ more than 6×3.

    A 7   **B** 3

    C 1   **D** 6

7 Which one is NOT a product of 4?

    A 40   **B** 24

    C 27   **D** 36

8 901¢ = ☐

    A $9.10   **B** $9.01

    C $901   **D** $0.91

9 _____ and one-seventh make 1 whole.

A $\frac{3}{7}$　　　　　　　　　　　　**B** $\frac{1}{7}$

C $\frac{5}{7}$　　　　　　　　　　　　**D** $\frac{6}{7}$

10 What fraction of the bar is shaded?

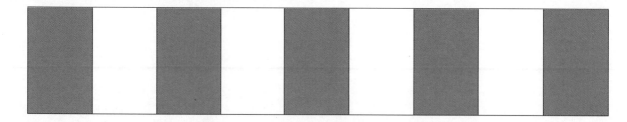

A $\frac{5}{9}$　　　　　　　　　　　　**B** $\frac{4}{5}$

C $\frac{1}{5}$　　　　　　　　　　　　**D** $\frac{4}{9}$

Section B (2 points each)

11 Seven hundreds, 3 ones, and 5 tens make _____.

12 $857 = 800 + \boxed{} + 7$

13 Write >, <, or = in the ◯.

$300 - 157 \bigcirc 230 - 157$

14 Complete the number pattern.

| 400 | | | 550 | 600 | | 700 |

15 One bag holds 4 hot dog buns.
How many hot dog buns are in 8 bags?

_____ hot dog buns are in 8 bags.

16 Kona puts 18 flowers equally into 3 vases.
How many flowers are in each vase?

_____ flowers are in each vase.

17 What is the total length of the two pencils?

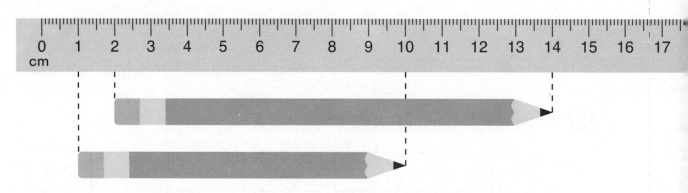

The total length is _____ cm.

18 Write the missing digits.

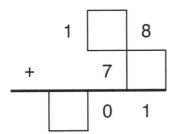

19 Carter has 2 five-dollar bills, 3 one-dollar bills and 2 quarters. How much money does he have?

He has $ _____.

20 Arrange the fractions in order, beginning with the smallest.

$\dfrac{1}{5}$ 　　 $\dfrac{1}{2}$ 　　 $\dfrac{1}{3}$ 　　 $\dfrac{1}{4}$

_____, _____, _____, _____

Section C (4 points each)

21 Amy saves $99.
Ben saves $9 more than Amy.

(a) How much does Ben save?

Ben saves $_____.

(b) How much do Amy and Ben save altogether?

Amy and Ben save $_____ altogether.

22 A pizza was cut into 10 equal slices.
Alex ate 1 slice, Emma ate 2 slices, and Dion ate 3 slices.

(a) Who ate $\frac{2}{10}$ of the pizza?

_____ ate $\frac{2}{10}$ of the pizza.

(b) What fraction of the pizza did they eat altogether?

They ate _____ of the pizza altogether.

23 Mr. Johnson has 4 bags of rice.
Altogether the rice weighs 24 lb.

(a) How much does one bag of rice weigh?

One bag of rice weighs _____ lb.

(b) How much do 10 bags of rice weigh?

10 bags of rice weigh _____ lb.

24 Jordan has $7.40.

(a) How much more money does she need to buy the dinosaur?

She needs $_____ more to buy the dinosaur.

(b) If she buys the car, how much money will she have left?

She will have $_____ left.

25 A farmer has 112 green bell peppers, 121 red bell peppers, and 56 yellow bell peppers.

(a) How many bell peppers does he have?

The farmer has _____ bell peppers.

(b) How many more red bell peppers does he have than yellow bell peppers?

He has _____ more red bell peppers than yellow bell peppers.

Test B

Continual Assessment 3

Section A (2 points each)

Circle the correct option: **A**, **B**, **C**, or **D**.

1 $997 - 99 =$ ☐

    A 900   **B** 897

    C 898   **D** 899

2 What is 80 more than 880?

    A 960   **B** 980

    C 888   **D** 800

3 $30 \div 3 >$ _____

A 3×7

B 4×6

C $40 \div 4$

D 3×3

4 What is the greatest number that can be made using 0, 3, and 9?

A 309

B 93

C 903

D 930

5 2 hundreds, 3 tens, and 15 ones make _____.

A 215

B 245

C 218

D 235

6 Which one can weigh about 10 kg?

A   a dog

B   a woman

C   a car

D   an orange

7 $0.03 is the same as _____.

A   30 cents

B   3 cents

C   300 cents

D   3 dollars

8 105 + 320 + 95 = ▭

A   420

B   510

C   520

D   500

9 There are _____ one-fifths in $\frac{4}{5}$.

A 5

B 1

C 4

D 9

10 Which one does NOT make 1 whole?

A $\frac{1}{7}$ and $\frac{5}{7}$

B $\frac{4}{5}$ and $\frac{1}{5}$

C $\frac{2}{8}$ and $\frac{6}{8}$

D $\frac{3}{4}$ and $\frac{1}{4}$

Section B (2 points each)

11 $97 + 97 =$ [ ]

12 _____ is 7 tens less than 5 hundreds and 2 tens.

13 Write >, <, or = in the ◯.

$9.25 − $1.00 ◯ $7.25 + $1.00

14 Complete the number pattern.

	344	443				839

15 Write >, <, or = in the ◯.

52 + 175 + 6 ◯ 22 + 195 + 6

16 Grace has a ribbon 9 times as long as this tape.
How long is her ribbon?

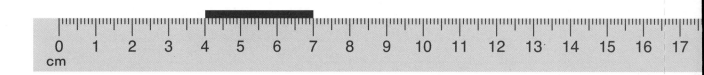

Her ribbon is _____ cm long.

17 Cody puts 20 muffins equally into some boxes.
There are 4 muffins in each box.
How many boxes of muffins are there?

There are _____ boxes of muffins.

18 Arrange the amounts of money in order from greatest to least.

| $5.05 • | $0.50 • | $5.50 • | $0.05 • | $55.00 • |

| | | | | |

19 Circle the fraction that is larger than $\frac{1}{6}$ and smaller than $\frac{1}{3}$.

$\frac{1}{7}$ $\frac{1}{2}$ $\frac{1}{5}$

20 Aaron saves $4 every day for 6 days.
Valentina saves $3 every day for 8 days.
Hunter saves $5 every day for 4 days.
Who saves the least amount of money?

_____ saves the least amount of money.

Section C (4 points each)

21 A bakery sold 200 muffins.
It sold 48 fewer scones than muffins.
It sold 29 more cookies than scones.

(a) How many scones did the bakery sell?

The bakery sold _____ scones.

(b) How many cookies did the bakery sell?

The bakery sold _____ cookies.

22 Han bought 4 packs of pens for $40.
There are 5 pens in each pack.

(a) How much does each pack of pens cost?

Each pack of pens costs $_____.

(b) How much does each pen cost?

Each pen costs $_____.

23 Mariam cuts a pie into 8 equal pieces.
She eats 1 piece of the pie.
She gives 3 pieces to her neighbors.

(a) What fraction of the pie does she give to her neighbors?

She gives _____ of the pie to her neighbors.

(b) What fraction of the pie does she have left?

She has _____ of the pie left.

24 A board game costs $9.99.
A pack of cards costs $8.74 less than the board game.

(a) How much does the pack of cards cost?

The pack of cards costs $_____.

(b) Carlos has 3 quarters and 2 dimes.
How much more money does he need to buy the pack of cards?

Carlos needs $_____ to buy the pack of cards.

25 A farmer has 39 mangoes.
She puts 4 mangoes in each bag.

(a) How many mangoes does she have left over?

She has _____ mangoes left over.

(b) She sells each bag of mangoes for $5.
How much does she sell all the bags of mangoes for?

She sells all of the bags of mangoes for $_____.

Name: _____

Date: _____

Test A

Chapter 12 Time

Section A (2 points each)

Circle the correct option: **A**, **B**, **C**, or **D**.

1 The clock shows _____.

A 7:12

B 7:21

C 8:12

D 7:22

2 The clock shows what time it is now.
What time was it 7 minutes ago?

A 8:48

B 9:50

C 9:48

D 9:07

3 The clock shows what time it is now.
In 3 hours, the hour hand will be between 4 and _____.

A 8

B 5

C 7

D 6

4 There are _____ hours in a day.

A 12

B 6

C 24

D 10

5 When the minute hand moves from 20 minutes to 3:00 to 10 minutes past 3:00, _____ minutes have passed.

A 15

B 10

C 60

D 30

Section B (2 points each)

6 What time is it?

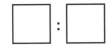

7 Draw the minute hand to show 18 minutes after 6.

8 What time will it be in 40 minutes?

It will be ☐ : ☐ in 40 minutes.

9 The clock is 11 minutes fast.
What is the actual time?

The actual time is ☐ : ☐ .

10 Write a.m. or p.m. in the blank.

Avery eats breakfast at 7:30 _____ .

Include a.m. or p.m. in all answers for questions 11–14.

11 Write the time for half past 3 in the afternoon.

_____

12 It is 1:15 a.m.

What time will it be in 12 hours?

It will be _____ in 12 hours.

13 Mei's swim lesson starts at 3:45 p.m.

The lesson is 50 minutes long.

What time does it end?

The lesson ends at _____.

14 Dion's class went on a two-hour field trip.
The field trip ended at 12:30 p.m.
What time did the field trip start?

The field trip started at _____.

15 Paul started making cookies at 5:35.
He finished making them at 6:20.
How many minutes did Paul spend making cookies?

Paul spent _____ minutes making cookies.

Name: _____

25 min   **Score**

Date: _____

| 30 |

Chapter 12 Time

Section A (2 points each)
Circle the correct option: **A**, **B**, **C**, or **D**.

1 How many minutes is it after one twenty?

A 5   **B** 7

C 2   **D** 27

2 The clock shows what time it is now.
What time will it be in 4 hours?

A 9:17   **B** 12:13

C 1:13   **D** 9:53

3 In another _____ minutes, it will be 10 o'clock.

A 22

B 20

C 38

D 10

4 There are _____ hours between noon and midnight.

A 24

B 10

C 12

D 2

5 Which clock cannot be correct?

A

B

C

D

Section B (2 points each)

6 Draw the minute hand to show five oh nine.

7 The clock shows what time it is now.
What was the time 55 minutes earlier?

The time was ☐ : ☐ .

8 Write a.m. or p.m. in the blanks.
Alex goes to bed at 8:30 _____ and gets up at 7:00 _____.

9 This clock is 7 minutes slow.
What is the actual time?

The actual time is ☐ : ☐ .

Include a.m. or p.m. in all answers for questions 10–15.

10 Write the time for 5 minutes after midnight.

_____

11 A math test started at 11:45 a.m.
The test was 25 minutes long.
What the time did the test end?

The test ended at _____.

12 Tom went to bed at 20 minutes past 10.
He got up 8 hours later.
What time did he get up?

Tom got up at _____.

13 A concert starts at 7:30 p.m.
The Dawson family arrives 38 minutes early.
What time did they arrive?

They arrived at _____.

14 A train ride from Jordan's house to the library takes 20 minutes.
Jordan gets off the train at 12:17 p.m. to go to the library.
What time did he get on the train?

He got on the train at _____.

15 Megan drove for 5 hours from Portland to Seattle.
She started at 8:10 a.m.
What time did she arrive in Seattle?

She arrived at _____.

Name: _____

Date: _____

Test A

Chapter 13 Capacity

Section A (2 points each)

Circle the correct option: **A**, **B**, **C**, or **D**.

1 Which container has the most water?

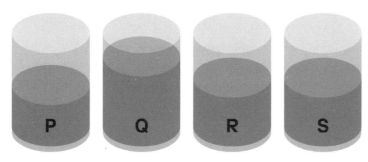

A Q   **B** R

C P   **D** S

2 A coffee pot can hold about _____ L of coffee.

A 200   **B** 20

C 50   **D** 2

3 There are 4 quarts in 1 gallon.
How many quarts are in 4 gallons?

A 16

B 6

C 8

D 12

4 Which of the following is likely to have a capacity of 300 L?

A a fish bowl

B a soup pot

C a bathtub

D a teacup

5 3 jugs each have the same amount of juice.
The total amount of juice in the jugs is 18 L.
How many liters of juice are in each jug?

A 21

B 6

C 9

D 3

Section B (2 points each)

Use the picture to answer questions 6–8.

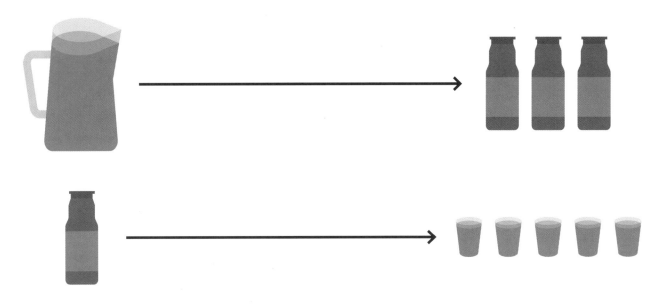

6 The pitcher can hold _____ bottles of water.

7 _____ cups of water will fill 1 bottle.

8 1 pitcher and 1 bottle hold _____ cups of water altogether.

9 A fish tank has a capacity of 48 gallons.
There are 39 gallons of water in it now.
How much more water is needed to fill the fish tank?

_____ gallons of water are needed to fill the fish tank.

10 A pot can hold 4 L of water.
How much water can 3 pots hold?

3 pots can hold _____ L of water.

Test B

Chapter 13 Capacity

Section A (2 points each)
Circle the correct option: **A**, **B**, **C**, or **D**.

1 How many cups of water can 3 of these pitchers hold?

A 7   **B** 24

C 16   **D** 8

2 Which container has a capacity of more than 1 L of water?

A a cup   **B** a water glass

C a mug   **D** a watering can

3 What is a reasonable capacity for a kid wading pool?

A 4 L

B 500 L

C 80 L

D 1 L

4 Lily pours 8 L of water equally into 4 bottles.
How many liters of water does she pour into each bottle?

A 4

B 2

C 8

D 12

5 There are 4 quarts in a gallon.
A fish tank has a capacity of 24 quarts.
How many gallons of water are needed to fill it?

A 8

B 12

C 4

D 6

Section B (2 points each)

Use the picture to answer questions 6–8.

6 Which has a greater capacity, the pitcher or the bucket? _____

7 _____ bottles of water are needed to fill the bucket.

8 The pitcher can hold _____ more mugs of water than the bottle.

9 A coffee shop had 25 gallons of milk.
The shop used some of the milk.
There are 17 gallons of milk left.
How much milk was used?

_____ gallons of milk were used.

10 A fish bowl can hold 11 L of water.
A fish tank can hold 45 L more water than the fish bowl.
What is the total capacity of the fish bowl and fish tank altogether?

The total capacity of the fish bowl and the fish tank is _____ L.

25 min **Score**

30

Test A

Chapter 14 Graphs

Section A (2 points each)
Circle the correct option: **A**, **B**, **C**, or **D**.

1 Each ⬤ stands for 3 cookies.

⬤ ⬤ ⬤ stands for _____ cookies.

A 1   **B** 9

C 3   **D** 4

2 ▪▪▪▪ stands for 16 boxes.

▪▪ stands for _____ boxes.

A 6   **B** 2

C 8   **D** 4

The picture graph shows the number of items sold by a coffee shop.
Use the graph to answer questions 3–5.

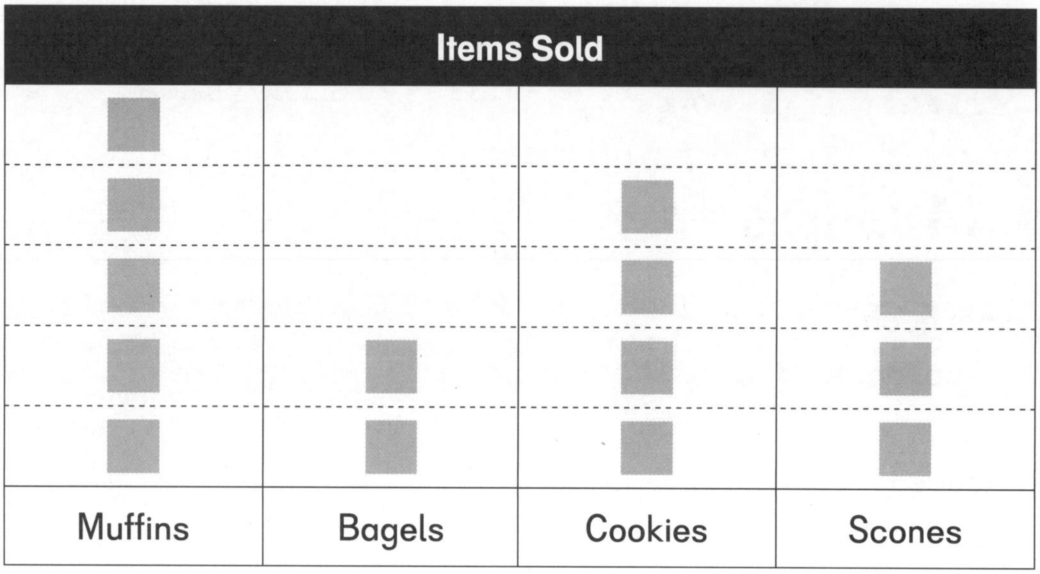

Items Sold			
▨			
▨		▨	
▨		▨	▨
▨	▨	▨	▨
▨	▨	▨	▨
Muffins	Bagels	Cookies	Scones

Each ▨ stands for 3 items.

3 The coffee shop sold _____ cookies.

A 12

B 4

C 6

D 8

4 The coffee shop sold _____ more muffins than scones.

A 2

B 4

C 6

D 9

5 _____ bagels and muffins were sold altogether.

A 14

B 7

C 9

D 21

Section B (2 points each)

This bar graph shows the number of pets in a pet shop.
Use the graph to answer questions 6—8.

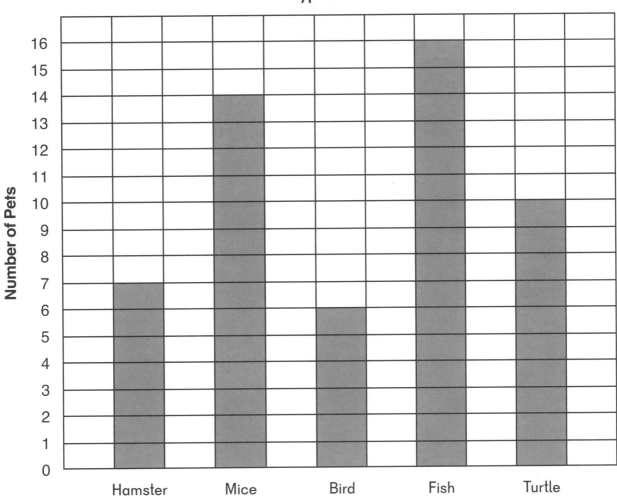

Types of Pets

6 The pet shop has _____ turtles.

7 The shop has fewer mice than _____.

8 The shop has 3 more turtles than _____.

The picture graph shows the amount of money each friend saved.
Use the graph to answer questions 9–11.

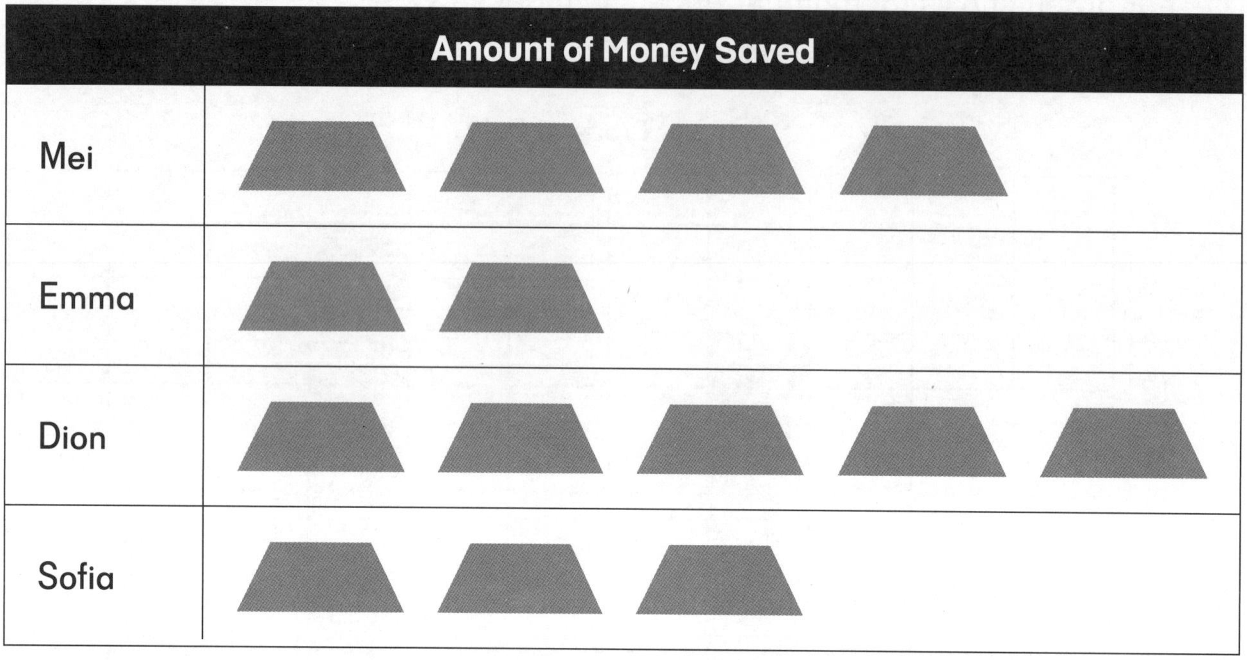

Amount of Money Saved	
Mei	
Emma	
Dion	
Sofia	

Each stands for 5 dollars.

9 Sofia saved $_____ less than Mei.

10 Emma, Dion, and Sofia saved $_____ altogether.

11 The amount of money that Alex saved is not shown on the graph.
He saved twice as much as Emma.
How much did Alex save?

Alex saved $_____.

The picture graph shows the number of seashells collected by some children.
Use the graph to answer questions 12–15.

Seashells Collected

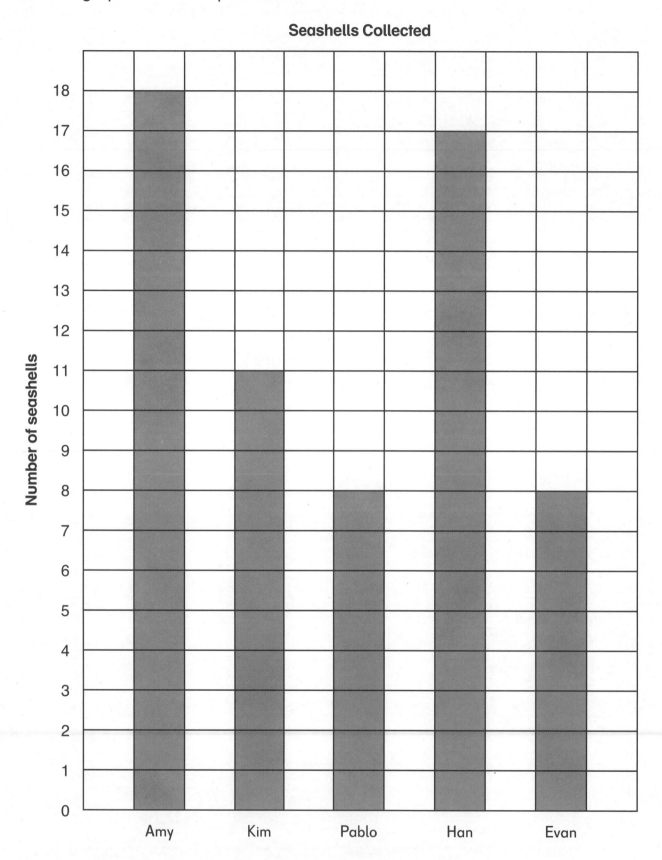

Chapter 14 Test A

12 _____ collected the most seashells.

13 _____ and _____ collected the same number of seashells.

14 Han collected _____ more shells than Kim.

15 The graph shows the number of seashells _____ children collected.

BLANK

Test B

Chapter 14 Graphs

Section A (2 points each)

Circle the correct option: **A**, **B**, **C**, or **D**.

1 ▲▲ stands for 10 trees.

▲▲▲▲ stands for _____ trees.

A 20

B 4

C 40

D 8

2 (bananas) stands for 10 monkeys.

(bananas) stands for _____ monkeys.

A 3

B 30

C 6

D 8

The picture graph shows the number of pieces of fruit used to make a fruit salad. Use the graph to answer questions 3–5.

Fruit Used				
		●		
		●		
	●	●		●
●	●	●		●
●	●	●	●	●
●	●	●	●	●
Apple	Peach	Strawberry	Orange	Banana

Each ● stands for 4 pieces of fruit.

3 There are _____ more strawberries than oranges in the salad.

    A 6   **B** 4

    C 8   **D** 16

4 _____ apples and bananas were used in the salad altogether.

    A 21   **B** 28

    C 14   **D** 2

5 _____ types of fruit are used to make the salad.

    A 6   **B** 7

    C 4   **D** 5

Section B (2 points each)

This bar graph shows the number of crayons in a tub.
Use the graph to answer questions 6—8.

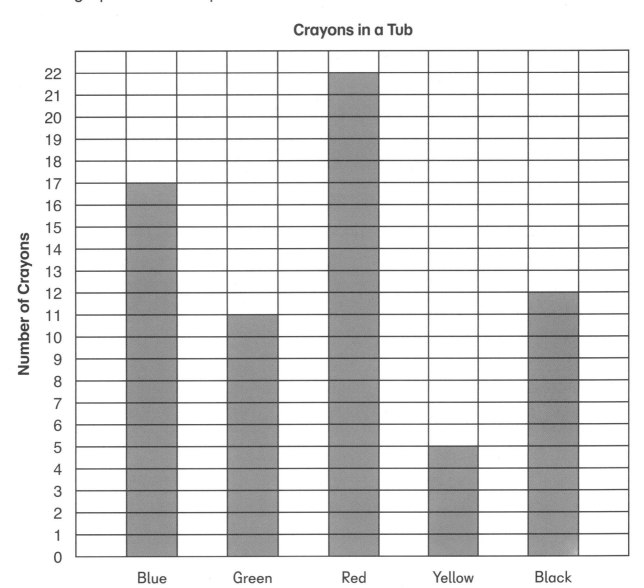

Crayons in a Tub

6 There are _____ green, yellow, and black crayons altogether.

7 There are the same number of blue and yellow crayons as _____ crayons.

8 The tub has the least number of _____ crayons.

The picture graph shows the number of books read by a group of friends during the summer.

Use the graph to answer questions 9–11.

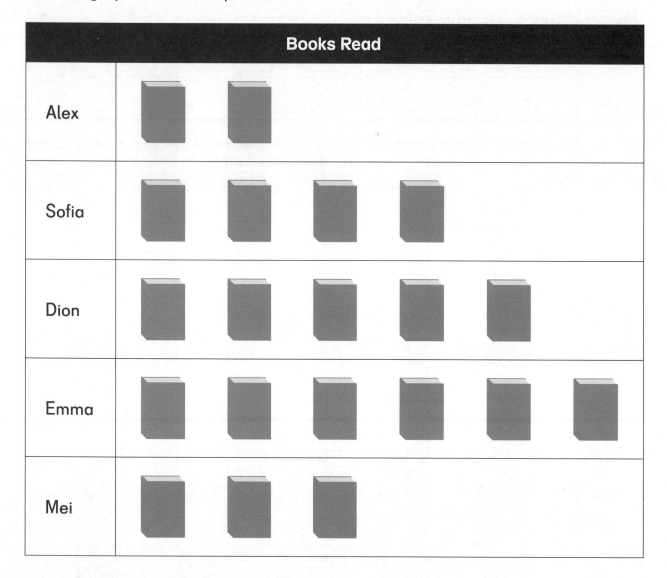

Each ▮ stands for 3 books.

9 _____ and _____ read fewer books than Sofia.

10 Alex and Dion read _____ books altogether.

11 Mei read _____ more books than Alex and _____ fewer books than Emma.

The picture graph shows the number of birthday cakes sold by a bakery each day of a certain week.

Use the graph to answer questions 12–15.

Birthday Cakes Sold

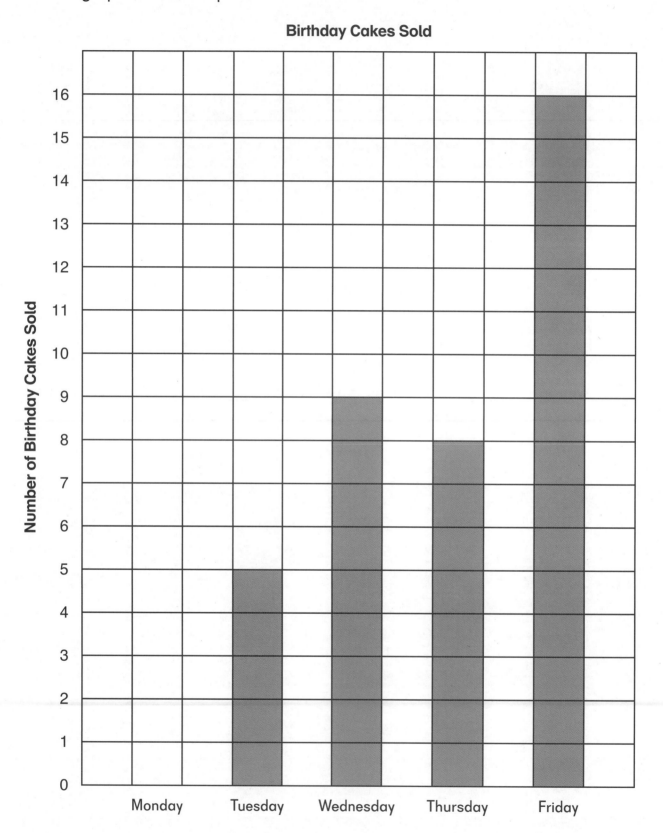

12 The bakery did not sell any birthday cakes on _____.

13 21 birthday cakes were sold altogether on _____ and _____.

14 The bakery sold _____ birthday cakes on Monday, Tuesday, and Thursday.

15 The number of birthday cakes the bakery sold on Saturday is not shown on the graph.
The bakery sold 10 more birthday cakes on Saturday than on Friday.
How many birthday cakes did the bakery sell on Saturday?

The bakery sold _____ birthday cakes on Saturday.

BLANK

Name: _____

Date: _____

25 min **Score**

30

Test A

Chapter 15 Shapes

Section A (2 points each)

Circle the correct option: **A**, **B**, **C**, or **D**.

1 Which one is a quadrilateral?

A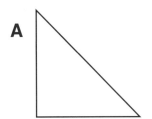

B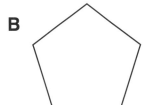

C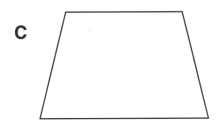

D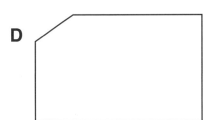

2 Which one is NOT a polygon?

A triangle

B pentagon

C rectangle

D circle

3 Which one does NOT have a curved surface?

A

B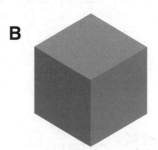

C

D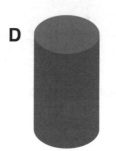

4 Which one has three sides and three corners?

A triangle

B square

C pentagon

D circle

5 Which shape comes next in the pattern?

?

A

B

C

D

Section B (2 points each)

6 Circle the figures that are closed shapes.

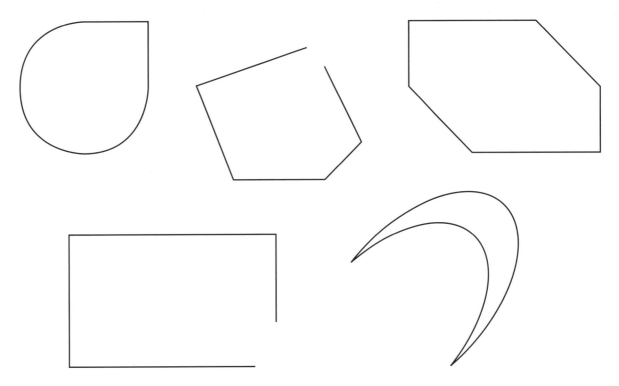

7 Draw straight lines to divide this figure into 3 triangles.

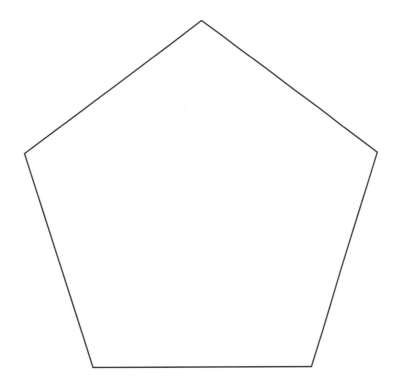

The above shape has _____ straight sides.

This shape has _____ sides and _____ corners.

Study the pattern.
Answer questions 10–12.

1st

10 Circle the missing shape.

11 The pattern repeats every _____ shapes.

12 Which of the following will be the 12th shape?
Circle it.

Circle the answers for questions 13–15.

13 It has 2 faces that are circles.
It has 1 curved surface.

cone   **cuboid**   **cylinder**   **sphere**

14 It has 8 corners.
Its flat surfaces are not all the same size.

sphere   **cuboid**   **cube**   **cone**

15 It can roll.
It has zero edges.

cylinder   **cone**   **sphere**   **cuboid**

25 min **Score**

30

Test B

Chapter 15 Shapes

Section A (2 points each)

Circle the correct option: **A**, **B**, **C**, or **D**.

1 How many of these shapes are quadrilaterals?

A 4   **B** 2

C 1   **D** 3

2 What is the 2nd shape of this pattern?

...   ...

5th   6th   7th   8th   9th

A   **B**

C   **D**

3 I am a shape and all of my faces are polygons.
What am I?

A sphere

B cone

C cylinder

D cuboid

4 A polygon with 15 corners has _____ sides.

A 30

B 10

C 15

D 5

5 How many edges will the 11th shape in this pattern have?

       ...

1st

A 2

B 8

C 1

D 0

6 Circle the shapes that are polygons.

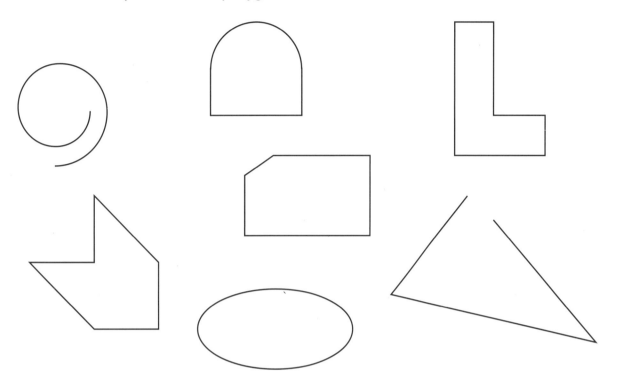

7 Draw straight lines to divide this figure into 2 triangles and 4 rectangles.

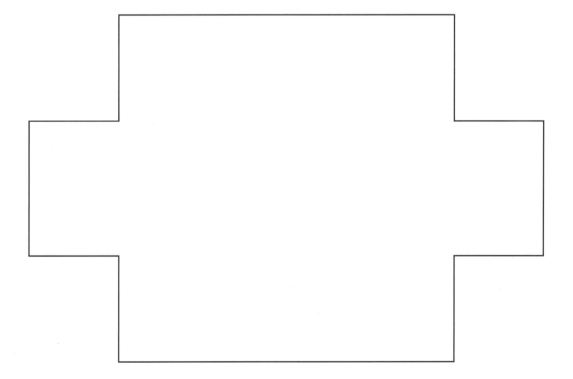

8 Check (✓) the statement that is true.

A triangle is a quadrilateral	
A triangle is a polygon	
A pentagon has 6 corners	

9 How many quarter-circles are in this shape?

_____

Study the pattern.
Answer questions 10–12.

1st

10 Circle what changes.

size   **shape**   **orientation**

11 Circle the 10th shape in this pattern.

12 Circle the 14th shape in this pattern.

Fill in the blanks for questions 13–15.

cube   cylinder

sphere   cuboid   cone

13 The _____, _____, and _____ have curved surfaces.

14 A _____ must have faces that are all squares.

15 The _____ and the _____ have both corners and straight edges.

Test A

Year-End Assessment

Section A (2 points each)
Circle the correct option: **A**, **B**, **C**, or **D**.

1 $800 - 98 =$ [ ]

A 700	**B** 702
C 798	**D** 102

2 4 hundreds, 0 tens, and 1 one make _____.

A 401	**B** 134
C 430	**D** 431

3 A paperclip is about 3 _____ long.

A ft	**B** m
C cm	**D** in

4 An octopus has 8 limbs.
How many limbs do 4 octopuses have?

A 16

B 24

C 40

D 32

5 $4.27 − $3.22 = ☐

A $1.25

B $4.05

C $1.05

D $7.49

6 ☐ + 7 + 200 = 277

A 70

B 200

C 77

D 7

7 25 ÷ 5 = ☐

A 30 ÷ 3

B 20 ÷ 4

C 4 × 2

D 40 ÷ 10

8 1 one-dollar bill, 5 quarters, and 2 nickels is _____.

A $2.45

B $2.10

C $2.35

D $1.35

9 There are _____ one-sixths in $\frac{5}{6}$.

A 6

B 3

C 1

D 5

10 What fraction of this shape is shaded?

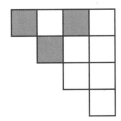

A $\frac{1}{3}$

B $\frac{3}{9}$

C $\frac{3}{7}$

D $\frac{3}{10}$

11 The farmers market opens at _____.

A 8:00 a.m.

B 10:00 p.m.

C 11:45 p.m.

D 1:00 a.m.

12 A fish tank holds less water than a _____.

A teapot   **B** bath tub

C cup   **D** water bottle

13 Which one is a polygon with 5 corners?

A    **B**

C 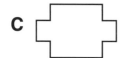   **D**

14 Which one has faces that are all polygons?

A   **B**

C   **D**

15 How many flat faces will the 10th shape in this pattern have?

        ...

1st

A 2   **B** 6

C 0   **D** 1

Section B (2 points each)

16 Circle the numbers with 4 in the tens place.

| 457 144 94 248 840 |

17 The clock shows what time it is now.

What time will it be in 8 minutes? _____

What time was it 2 hours ago? _____

18 Write >, <, or = in the ◯.

$7.08 ◯ 708¢

19 Circle the numbers that are less than 298 and greater than 98.

| 198 89 308 99 297 |

20 A farmer sold 386 kg of yellow potatoes.
He sold 205 kg of red potatoes.
How many kilograms of potatoes did he sell altogether?

He sold _____ kg of potatoes altogether.

21 Alex cuts a 16 ft tape into 4 equal pieces.
How long is each piece?

Each piece is _____ ft long.

22 Laila wants to put 3 tacos on each plate.
She has 9 plates.
How many tacos does she need?

She needs _____ tacos.

Use the picture to answer questions 23–24.

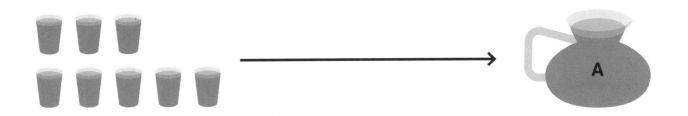 is 1 unit.

23 Container C can hold 2 more units than Container _____.

24 14 units of water can fill Containers _____ and _____.

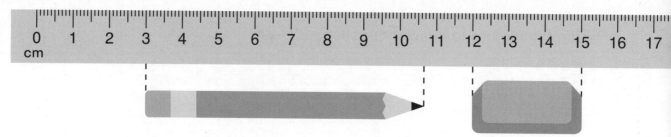

25

The total length of the pencil and the eraser is about _____ cm.

26 It takes 25 minutes to cook a pot of rice.
Rita started cooking a pot of rice at 5:55 p.m.
What time will the rice be done?

The rice will be done at _____.

27 Write >, <, or = in the ◯.

$\frac{1}{2}$ ◯ $\frac{1}{3}$

The picture graph shows the number of flowers sold by a shop.
Use the graph to answer questions 28–30.

Number of flowers sold	
Monday	🌷🌷🌷🌷🌷🌷🌷🌷
Tuesday	🌷🌷🌷🌷🌷
Wednesday	🌷🌷🌷🌷🌷🌷🌷
Thursday	🌷🌷🌷🌷🌷🌷
Friday	🌷🌷🌷🌷🌷🌷🌷🌷🌷

Each 🌷 stands for 10 flowers.

28 On _____ and _____, the shop sold the same number of flowers.

29 The shop sold _____ more flowers on Friday than on Tuesday.

30 The shop sold _____ flowers on Monday, Tuesday, and Thursday.

Section C (4 points each)

31 A farmer sold 24 gallons of cow's milk.
She sold 19 fewer gallons of goat's milk than cow's milk.

(a) How many gallons of goat's milk did she sell?

She sold _____ gallons of goat's milk.

(b) She sold the goat's milk for $8 per gallon.
How much did she sell the goat's milk for in total?

She sold the goat's milk for $_____ in total.

32 Each box of crayons costs $4.

(a) How much do 3 boxes of crayon cost?

Three boxes of crayons cost $ _____.

(b) Jason has $36.
How many boxes of crayons can he buy?

He can buy _____ boxes of crayons.

33 A box of toy cars has 105 red cars, 75 blue cars, and 22 white cars. How many toy cars are there in the box altogether?

There are _____ toy cars altogether.

34 The clocks show when Emma's swimming lesson begins and ends.

Begin   End

(a) What time does Emma's lesson begin?

The lesson begins at _____.

(b) How long is the swimming lesson?

The lesson is _____ minutes long.

35 381 people were at a baseball game on Saturday.
99 fewer people were at the game on Sunday than on Saturday.

(a) How many people were at the game on Sunday?

_____ people were at the game on Sunday.

(b) How many people were at the game on Saturday and Sunday altogether?

_____ people were at the game on Saturday and Sunday altogether.

BLANK

Test B

Year-End Assessment

Section A (2 points each)
Circle the correct option: **A**, **B**, **C**, or **D**.

1 234 is _____ tens less than 2 hundreds, 3 tens, and 24 ones

A 0   **B** 4

C 2   **D** 20

2 The digit 3 in _____ has a value of 30.

A 381   **B** 123

C 303   **D** 138

3 $\boxed{} - 99 = 585$

A 485   **B** 685

C 684   **D** 486

4 $10 \times 4 <$ ☐

    A 9×5 **B** $15 \div 3$

    C $40 \div 4$ **D** 8×3

5 What is the least number that can be made using 8, 9, and 1?

    A 981   **B** 189

    C 891   **D** 198

6 5×7 is _____ more than 3×7.

    A 7   **B** 14

    C 21   **D** 2

7 A watermelon weighs about _____.

    A 40 kg   **B** 1 g

    C 20 lb   **D** 1 lb

8 1 five-dollar bill, 7 quarters, and 20 pennies is _____.

    A $5.95   **B** $6.97

    C $2.95   **D** $6.95

9 $97 + 542 + 138 = \boxed{}$

    A 677   **B** 777

    C 767   **D** 667

10 $\$8.00 - 85¢ = \boxed{}$

    A $1.15   **B** $8.85

    C $7.15   **D** $0.15

11 How many curved surfaces does the shape that is between the sphere and the cuboid have?

    A 0   **B** 1

    C 2   **D** 6

12 15 minutes before noon is the same time as _____.

   A 12:15 p.m.   **B** 11:45 a.m.

   C 11:45 p.m.   **D** 12:15 a.m.

13 This figure is _____.

   A a closed shape   **B** a pentagon

   C a polygon   **D** an open shape

14 What is the 1st shape in this pattern?

8th

   A   **B**

   C   **D**

15 A chicken pot pie costs $3.
How many pot pies can Jada buy with $23?

   A 7   **B** 23

   C 21   **D** 10

Section B (2 points each)

16 Write 213 in words.

_____

17 Write the numbers in order from greatest to least.

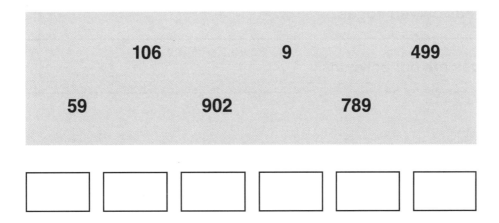

106		9		499
59	902		789	

18 Complete the number pattern.

360			510	560		660

19 Circle the fractions that are smaller than $\frac{1}{4}$.

$\frac{1}{3}$ $\frac{1}{5}$ $\frac{1}{2}$ $\frac{1}{8}$ $\frac{1}{6}$

20 Write >, <, or = in the ◯.

278 − 98 ◯ 97 − 85

21 Check (✓) the statement that is NOT true.

A quadrilateral is a polygon.	
A polygon has curved edges.	
Open shapes are not polygons.	

22 Draw straight lines to divide this figure into 3 triangles.

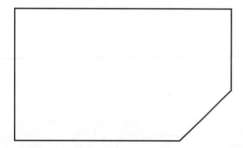

23 Dion gets off the school bus at 2:35 p.m.
Emma gets off the bus 18 minutes later.
What time does Emma get off the bus?

Emma gets off the bus at _____.

Use the pictures to answer questions 24—25.

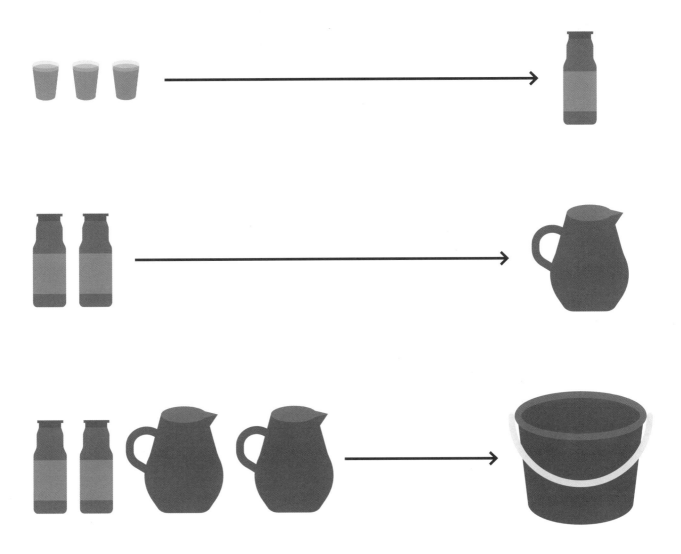

24 _____ bottles of water are needed to fill the bucket.

25 _____ cups of water are needed to fill the bucket.

The picture graph shows the amount of money each friend saved.
Use the graph to answer questions 26–27.

Money Saved				
			★	
★			★	
★	★		★	★
★	★	★	★	★
★	★	★	★	★
Alex	Dion	Emma	Mei	Sofia

26 Emma, Mei, and Sofia saved $30 altogether.
How much does each ★ stand for?
Each ★ stands for $_____.

27 Dion and Sofia together saved $_____ more than Alex.

The bar graph shows the number of each kind of pasta shape Sam chose to use for an artwork.

Use the graph to answer questions 28–30.

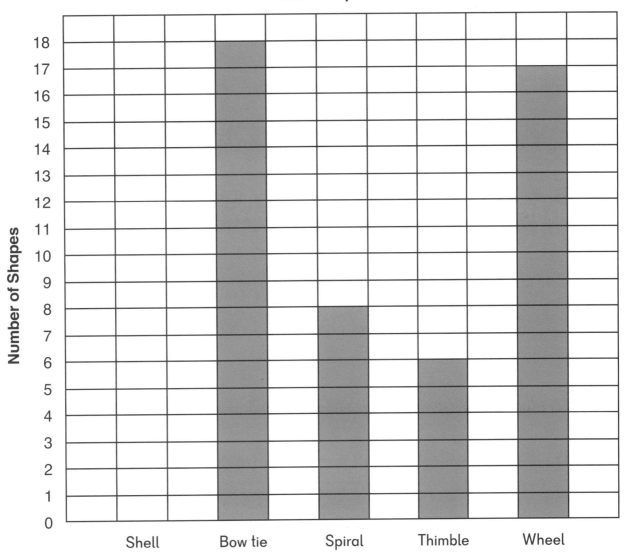

Pasta Shapes Used

28 The shape that Sam used most was the _____.

29 He did not use any _____.

30 He used _____ spirals, thimbles, and wheels altogether.

Section C (4 points each)

31 The distance from Kalama's home to the school is 467 m.
The distance from Matt's home to the school is 476 m.

(a) How much closer is Kalama's home than Matt's home to the school?

Kalama's home is _____ m closer than Matt's home.

(b) Kalama jogged from her home to school and then to Matt's home.
What is the total distance that she jogged from her home to Matt's home?

She jogged _____ m from her home to Matt's home.

32 336 children and 199 adults visited a museum last week.

(a) How many people visited the museum last week?

_____ people visited the museum last week.

(b) How many more children than adults visited the museum?

_____ more children than adults visited the museum.

33 Jason cut a cake into 10 equal pieces.
He and his cousins ate 8 pieces of the cake altogether.

(a) What fraction of the cake was each piece of cake?

Each piece of cake was _____ of the cake.

(b) What fraction of the cake was left?

_____ of the cake was left.

34 Kai's piano lesson started at 12:15 p.m.
The lesson was 50 minutes long.
He arrived 20 minutes early for his lesson.

(a) What time did he arrive?

He arrived at _____.

(b) His lesson started on time.
What time did the lesson end?

The lesson ended at _____.

35 John bought 35 burritos.

He put 7 burritos in the refrigerator.

He put the rest of the burritos equally on 4 plates.

How many burritos are on each plate?

There are _____ burritos on each plate.

Answer Key

Chapter 8 Mental Calculation

1 D

2 A

3 D

4 C

5 B

6 136

7 698

8 250

9 651

10 899

11

219	227	235
243	251	259

12 =

13 63

14 103

15 19

Chapter 8 Mental Calculation

1 B

2 D

3 B

4 C

5 D

6 998

7 3

8 53

9 98

10 960

11 <

12

99	198	297
396	495	594

13 573

14 82

15 935

Chapter 9 Multiplication and Division of 3 and 4

1 B

2 D

3 A

4 D

5 C

6 9

7 25 (18) (30) 20 8 (9)

8 ÷

9 18

10 4

11 24

12 3

13 36

14 7

15 (a) 10

(b) 24

Chapter 9 Multiplication and Division of 3 and 4

1 B

2 A

3 C

4 B

5 C

6 3

7 =

8 12, 32

9

| 9 | 20 | 16 | 21 | (24) | 30 |

10 7

11 4

12 7

13 (a) 6

(b) 2

14 (a) 21

(b) 28

15 2

Chapter 10 Money

1 B

2 C

3 A

4 D

5 B

6 86

7 13

8 >

9 | $0.28 | | $2.08 | | $2.80 |

| $8.00 |

10 9.10

11 3.20

12 2.05

13 5.80

14 4.54

15 (a) 3.80

(b) 6.92

Chapter 10 Money

1 B

2 D

3 B

4 C

5 A

6 4.00

7 =

8 | $3.83 $4.73 ($3.63) $3.93 |

9 5.50

10 9.80

11 dinosaur

12 monkey

13 (a) 1.65

    (b) 4.70

14 (a) 4.70

    (b) 8.65

15 1.01

Chapter 11 Fractions

1 C

2 B

3 D

4 C

5 A

6 Answers may vary.

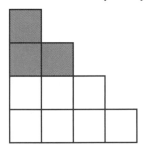

7 $\frac{4}{7}$

8 $\frac{7}{12}$

9 Answers may vary.

10 $\frac{1}{3}$

11 $\frac{1}{8}$

12 $\frac{1}{7}, \frac{1}{4}, \frac{1}{2}$

13 Carlos

14 $\frac{5}{8}$

15 $\frac{7}{9}$

Chapter 11 Fractions

1 C

2 B

3 D

4 A

5 C

6 Answers may vary.

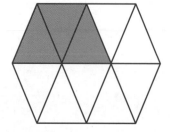

7 $\frac{2}{7}$, $\frac{5}{7}$

8 $\frac{8}{9}$

9 $\frac{1}{6}$

10 $\frac{5}{10}$

11 $\frac{1}{5}$, $\frac{1}{7}$, $\frac{1}{8}$, $\frac{1}{10}$

12 $\frac{1}{2}$

13 morning

14 $\frac{4}{10}$

15 blue

Continual Assessment 3

1 B

2 C

3 A

4 C

5 D

6 B

7 C

8 B

9 D

10 A

11 753

12 50

13 >

14

400	**450**	500
550	600	**650**

700

15 32

16 6

17 21

18

	1	**2**	8
+		7	**3**
	2	0	1

19 13.50

20 $\frac{1}{5}, \frac{1}{4}, \frac{1}{3}, \frac{1}{2}$

21 (a) 108

(b) 207

Continual Assessment 3

22 (a) Emma

(b) $\frac{6}{10}$

---

23 (a) 6

(b) 60

---

24 (a) 2.45

(b) 1.65

---

25 (a) 289

(b) 65

Continual Assessment 3

1 C

2 A

3 D

4 D

5 B

6 A

7 B

8 C

9 C

10 A

11 194

12 450

13 =

14

245	344	443
542	641	740
839		

15 >

16 27

17 5

18

$55.00	$5.50	$5.05
$0.50	$0.05	

19 $\frac{1}{5}$

20 Hunter

21 (a) 152

(b) 181

Continual Assessment 3

22 (a) 10

(b) 2

---

23 (a) $\frac{3}{8}$

(b) $\frac{4}{8}$

---

24 (a) 1.25

(b) 0.30

---

25 (a) 3

(b) 45

Chapter 12 Time

1 A

2 C

3 B

4 C

5 D

6 11:03

7

8 5:05

9 2:44

10 a.m.

11 3:30 p.m.

12 1:15 p.m.

13 4:35 p.m.

14 10:30 a.m.

15 45

Chapter 12 Time

1 B

2 C

3 A

4 C

5 C

6

7 2:20

8 p.m., a.m.

9 4:12

10 12:05 a.m.

11 12:10 p.m.

12 6:20 a.m.

13 6:52 p.m.

14 11:57 a.m.

15 1:10 p.m.

Chapter 13 Capacity

1 A

2 D

3 A

4 C

5 B

6 3

7 5

8 20

9 9

10 12

Chapter 13 Capacity

1 B

2 D

3 C

4 B

5 D

6 bucket

7 6

8 2

9 8

10 67

Chapter 14 Graphs

1 B

2 C

3 A

4 C

5 D

6 10

7 fish

8 hamsters

9 5

10 50

11 20

12 Amy

13 Pablo, Evan

14 6

15 5

Chapter 14 Graphs

1 A

2 C

3 D

4 B

5 D

6 28

7 red

8 yellow

9 Alex, Mei

10 21

11 3, 9

12 Monday

13 Tuesday, Friday

14 13

15 26

Chapter 15 Shapes

1 C

2 D

3 B

4 A

5 B

6

7 Answers may vary.

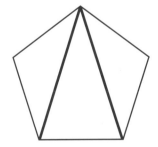

8 3

9 5, 5

10

11 4

12

13 cylinder

14 cuboid

15 sphere

Chapter 15 Shapes

1 B

2 C

3 D

4 C

5 D

6

7 Answers may vary.

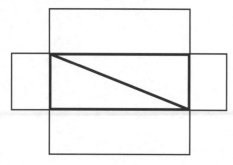

8

A triangle is a quadrilateral	
A triangle is a polygon	✓
A pentagon has 6 corners	

9 6

10 (size) shape (orientation)

11

12

13 sphere, cylinder, cone

14 cube

15 cube, cuboid

Year-End Assessment

1 B

2 A

3 C

4 D

5 C

6 A

7 B

8 C

9 D

10 D

11 A

12 B

13 D

14 C

15 A

16 457 (144) 94 (248) (840)

17 12:23, 10:15

18 =

19 (198) 89 308 (99) (297)

20 591

21 4

22 27

23 A

24 A, B

Year-End Assessment

25 11

26 6:20 p.m.

27 >

28 Monday, Wednesday

29 40

30 210

31 (a) 5

(b) 40

32 (a) 12

(b) 9

33 202

34 (a) 11:40 a.m.

(b) 55

35 (a) 282

(b) 663

Year-End Assessment

1 C

2 D

3 C

4 A

5 B

6 B

7 C

8 D

9 B

10 C

11 A

12 B

13 D

14 C

15 A

16 Two hundred thirteen

17

902	789	499
106	59	9

18

360	**410**	**460**
510	560	**610**

660

19 $\frac{1}{3}$ $\boxed{\frac{1}{5}}$ $\frac{1}{2}$ $\boxed{\frac{1}{8}}$ $\boxed{\frac{1}{6}}$

20 >

21

A quadrilateral is a polygon	
A polygon has curved edges	✓
Open shapes are not polygons	

Year-End Assessment

22 Answers may vary.

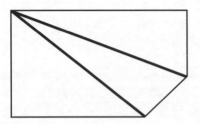

23 2:53 p.m.

24 6

25 18

26 3

27 6

28 bow tie

29 shells

30 31

31 (a) 9

(b) 943

32 (a) 535

(b) 137

33 (a) $\frac{1}{10}$

(b) $\frac{2}{10}$

34 (a) 11:55 a.m.

(b) 1:05 p.m.

35 7